PowerKids Readers:

ROAD MACHINES™

Road Scrapers

Joanne Randolph

The Rosen Publishing Group's
PowerKids Press™
New York

1

For Joseph Hobson, with love

Published in 2002 by The Rosen Publishing Group, Inc.
29 East 21st Street, New York, NY 10010

First Edition

Book Design: Michael Donnellan

Photo Credits: p. 5 © Photodisc; pp. 7, 11, 15, 17 © Highway Images/Genat; pp. 9, 13, 19, 21 © Corbis.

Randolph, Joanne.
  Road scrapers / Joanne Randolph.
    p. cm. — (Road machines)
  ISBN 0-8239-6042-0 (library binding)
    1. Road scrapers—Juvenile literature. [1. Road machinery.]   I. Title.
TE223 .R38 2002
625.7'5—dc21
                                                      2001000656

Manufactured in the United States of America

Contents

This is a scraper.

5

Scrapers remove lots of dirt. They remove the dirt so roads can be built.

Some scrapers need to be towed by tractors.

9

A scraper has blades that can be raised or lowered. The blades help remove the dirt.

A scraper has a bowl to carry the dirt. The dirt in the bowl can be dumped out when the bowl gets full.

13

This scraper is making the path for a new road. It drives back and forth until the path is deep enough to be paved.

15

A person drives the scraper. The driver decides how deep the blades need to dig. The driver dumps the dirt where it is needed.

Scrapers also help build parking lots and buildings. They help make the land flat.

Scrapers do a lot of work.

Words to Know

bowl

scraper

tractor

Here is another book to read about scrapers:
Diggers and other Construction Machines
(Machines at Work)
by Jon Richards
Copper Beech Books

To learn more about scrapers, check out this
Web site:
www.field-guides.com/html/scrapers.html

Index

Word Count: 133

Note to Librarians, Teachers, and Parents

PowerKids Readers are specially designed to help emergent and beginning readers build their skills in reading for information. Simple vocabulary and concepts are paired with photographs of real kids in real-life situations or stunning, detailed images from the natural world around them. Readers will respond to written language by linking meaning with their own everyday experiences and observations. Sentences are short and simple, employing a basic vocabulary of sight words, as well as new words that describe objects or processes that take place in the natural world. Large type, clean design, and photographs corresponding directly to the text all help children to decipher meaning. Features such as a contents page, picture glossary, and index help children get the most out of PowerKids Readers. They also introduce children to the basic elements of a book, which they will encounter in their future reading experiences. Lists of related books and Web sites encourage kids to explore other sources and to continue the process of learning.